D0743171

PLANT POWER
PHOTOSYNTHESIS

by Karen Latchana Kenney

Ideas for Parents and Teachers

Pogo Books let children practice reading informational text while introducing them to nonfiction features such as headings, labels, sidebars, maps, and diagrams, as well as a table of contents, glossary, and index.

Carefully leveled text with a strong photo match offers early fluent readers the support they need to succeed.

Before Reading

• "Walk" through the book and point out the various nonfiction features. Ask the student what purpose each feature serves.

• Look at the glossary together. Read and discuss the words.

Read the Book

• Have the child read the book independently.

• Invite him or her to list questions that arise from reading.

After Reading

• Discuss the child's questions. Talk about how he or she might find answers to those questions.

• Prompt the child to think more. Ask: What did you know about photosynthesis before reading this book? After reading it, what more would you like to learn?

Pogo Books are published by Jump!
5357 Penn Avenue South
Minneapolis, MN 55419
www.jumplibrary.com

Library of Congress Cataloging-in-Publication Data

Names: Kenney, Karen Latchana, author.
Title: Photosynthesis / by Karen Latchana Kenney.
Description: Minneapolis, MN : Jump!, Inc., [2018]
Series: Plant power
Audience: Ages 7-10."Pogo Books."
Includes bibliographical references and index.
Identifiers: LCCN 2018001489 (print)
LCCN 2018008506 (ebook)
ISBN 9781624968822 (ebook)
ISBN 9781624968808 (hardcover : alk. paper)
ISBN 9781624968815 (paperback)
Subjects: LCSH: Photosynthesis–Juvenile literature.
Classification: LCC QK882 (ebook)
LCC QK882 .K39 2018 (print) | DDC 572/.46–dc23
LC record available at https://lccn.loc.gov/2018001489

Editor: Jenna Trnka
Book Designer: Molly Ballanger

Photo Credits: Subbotina Anna/Shutterstock, cover; Iurii Kachkovskyi/Shutterstock, 1 (leaf); David Gilder/Shutterstock, 1 (magnifying glass); Sina Jasteh/Shutterstock, 1 (veins); jannoon028/Shutterstock, 3; Manfred Ruckszio/Shutterstock, 4; imageBROKER/SuperStock, 5; Juergen Faelchle/Shutterstock, 6-7; Nataliya Hora/Shutterstock, 8-9; Fotofermer/Shutterstock, 10; Cornel Constantin/Shutterstock, 11; ThomasVogel/iStock, 12-13; Miki Studio/Shutterstock, 14-15; Andrew Burgess/Shutterstock, 16; Casther/Shutterstock, 17; Adriana Margarita Larios Arellano/Shutterstock, 18-19; Beau Lark/Corbis/VCG/Getty, 20-21; shutterdk/Shutterstock, 23.

Printed in the United States of America at Corporate Graphics in North Mankato, Minnesota.

TABLE OF CONTENTS

CHAPTER 1
LIGHT CATCHERS

A beech tree's **bud** twists and bends. It begins to open. Inside are leaves folded up tightly.

bud

The leaves slowly stretch out and unfold. They spread wide and flat. They must get as big as possible. They are the tree's light catchers.

Beams of warm sunlight fall onto a plant's leaves. This light is made of **energy**. Sunlight is an important ingredient. Plants need it to make food. This is photosynthesis.

DID YOU KNOW?

Sunlight travels far to reach plants. This energy travels 93 million miles (150 million kilometers)! And fast. It only takes around eight minutes to zoom to Earth!

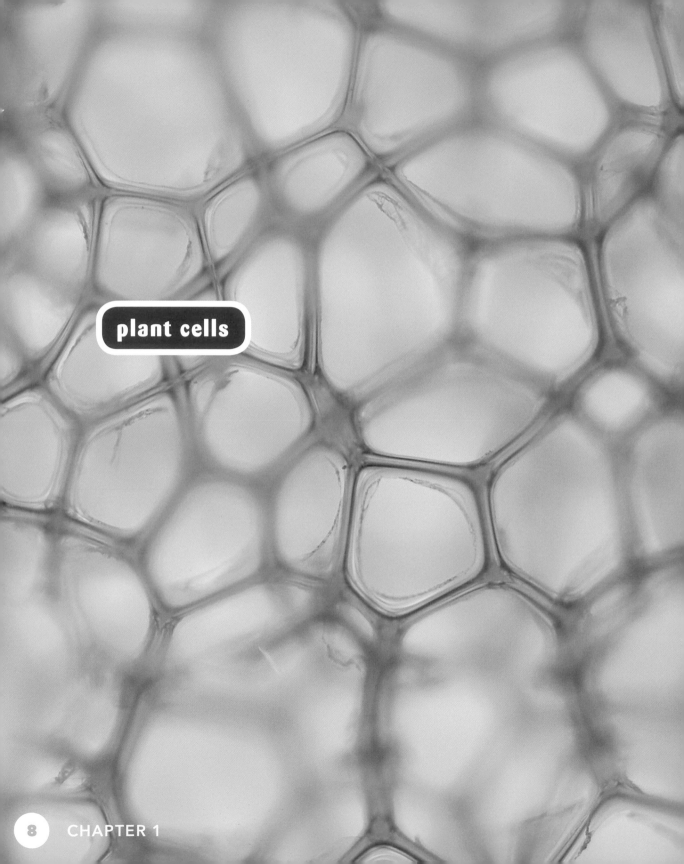

plant cells

Once sunlight hits the leaves, it is trapped. How? Special **cells** contain **chlorophyll**. It **absorbs** the sunlight's energy. Now the plant just needs a few more ingredients.

DID YOU KNOW?

Some plants make food even when their leaves fall off in winter. The quaking aspen tree has chlorophyll in its bark. It can make food all winter long.

CHAPTER 2

PORES AND ROOTS

vein

Look closely at a leaf. It is not just flat and smooth. It is covered with a maze of **veins**.

stomata · · · · ▶

Look even closer. You'll see **pores** called **stomata**. They open and close. The pores let in ingredients for photosynthesis. **Carbon dioxide** is a gas in the air. Water is in the air, too. They flow into a leaf through its pores.

Water is also in the soil. A plant's **roots** reach deep into the soil. The roots soak up water and **nutrients**. They travel up through the stem. They move through veins into the leaf. Then water and nutrients reach the cells.

DID YOU KNOW?

Roots also hold a plant in the soil. Some plants have a large **taproot**. Other plants have many thin roots.

roots ·····▶

Now the leaf's cells have the right ingredients to make food. Energy from the sun begins the process. The energy moves through the cells. It changes as it reacts with carbon dioxide and water. It becomes a kind of sugar that the hungry plant uses to grow.

TAKE A LOOK!

Different parts of a plant gather the ingredients for its food.

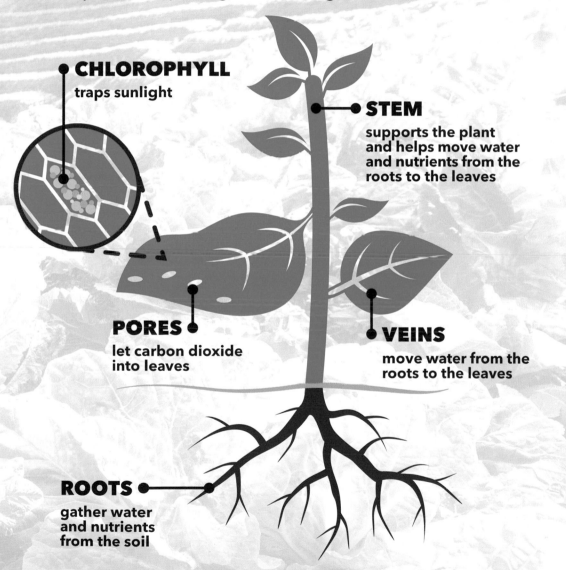

CHLOROPHYLL
traps sunlight

STEM
supports the plant
and helps move water
and nutrients from the
roots to the leaves

PORES
let carbon dioxide
into leaves

VEINS
move water from the
roots to the leaves

ROOTS
gather water
and nutrients
from the soil

CHAPTER 3

GIVING PLANTS

Plants grow bigger and bigger using the sugar they make. New buds form. Leaves unfold. Plants grow taller. They reach up toward the sun.

But plants don't use all of their food right away. They store some in their roots. The roots grow large. Some become sweet and full of nutrients. They are the vegetables we eat.

Apples taste sweet. That is because a plant's food fuels growing fruit. Sweet fruit is a great package for the seeds inside. Many animals eat fruit. They carry the seeds in their bodies. With their waste, the animals leave the seeds in new places. Soon **seedlings** will grow.

When plants make their food, they make something else, too. They release **oxygen** from their pores. It goes into the air and helps all creatures on Earth. People and animals breathe oxygen in and use it to make energy in their bodies. Photosynthesis helps plants grow. And it also helps keep the planet healthy.

ACTIVITIES & TOOLS

BREATHING LEAVES

You can't see the gases moving in and out of a leaf on a plant. But try this activity to see how leaves "breathe."

What You Need:
- large glass bowl
- warm water
- leaf
- rock

❶ Fill the glass bowl with warm water.

❷ Go outside and pick a leaf off a plant or tree.

❸ Place the leaf in the bowl. Weigh it down with the rock so that it stays underwater.

❹ Put the bowl in a sunny spot. Then wait a few hours before checking on the leaf.

❺ Lift up the rock. Check on the leaf. What do you see? Small bubbles are on the leaf's surface and at the sides of the bowl. This shows the oxygen that the leaf releases during photosynthesis.

GLOSSARY

absorbs: Takes in or soaks up.

bud: A small knob on a plant that grows into a leaf, shoot, or flower.

carbon dioxide: A gas that plants absorb.

cells: Basic, small parts of plants or animals.

chlorophyll: The green substance in plants that gathers energy from the sun.

energy: The ability to do work.

nutrients: Proteins, minerals, and other substances plants need to grow and live.

oxygen: A gas that plants release into the air.

pores: Tiny holes.

roots: The parts of plants that grow underground and absorb nutrients and water.

seedlings: Young plants that grow from seeds.

stomata: Pores in leaves that open and close.

taproot: A large, central root of a plant.

veins: Small tubes in a leaf that carry water and nutrients.

INDEX

TO LEARN MORE

Learning more is as easy as 1, 2, 3.

1) Go to www.factsurfer.com

2) Enter "photosynthesis" into the search box.

3) Click the "Surf" button to see a list of websites.

With factsurfer, finding more information is just a click away.